THE BOBBS-MERRILL STUDIES IN SOCIOLOGY

# Science: Some Sociological Perspectives

## Nicholas C. Mullins

THE BOBBS-MERRILL COMPANY, INC.
INDIANAPOLIS • NEW YORK

Copyright © 1973 by The Bobbs-Merrill Company Inc.
Printed in the United States of America
First Printing

**Library of Congress Cataloging in Publication Data**
Mullins, Nicholas C
Science: some sociological perspectives.
(Bobbs-Merrill studies in sociology)
Bibliography: p.
1. Science—Social aspects—United States.
I. Title.
Q175.5.M86      301.24'3      72–12826
ISBN 0–672–61205–4 (pbk)

# Introduction

We live in a scientific age. Students in most educational situations study something called "science." Advertisements trumpet a "new scientific discovery" and show a man in a white lab coat (a potent symbol of science) extolling the virtues of some new way to stop pain three different ways. Congressmen vote money for science, watch space shots, break ground for laboratory buildings, and invite streams of ecologists, demographers, economists, physicists, and others to testify on the scientific aspects of proposed legislation. To evaluate science itself, we fill our bookshelves with descriptions of the dehumanizing and monstrous, or glorious and wondrous, effects of science.

In these and more substantial ways, science is a major part of the modern world and in particular of industrial societies. The United States alone (the world's most industrialized society) has more than 750,000 people with scientific training and employment as researchers, engineers, teachers, and so on, of science (OECD, 1968:45). This number represents the largest number of science-associated persons ever assembled in one country. Further, it represents 45 percent of all scientists who have ever existed in the history of the world.[1] In science, however, unlike industry, an increased number of workers does not necessarily result in greater production except as production can be measured by scientific papers, more laboratories, and more students.

Science results from the constant battle by groups of scientists to understand and explain experiences according to a conceptual scheme shared among them. For example, the chemist who discovered that xenon combines with other elements such as fluorine found that he and his colleagues had no way to include that fact in their conceptual scheme. So they revised that scheme to encompass the new finding. By contrast, discovery of the replicating mechanism for life, the splitting and rebuilding of DNA molecules, which might be seen as more startling, was easily accepted by chemists because the rebuilding mechanism, the familiar hydrogen bond, was already an accepted part of chemistry's conceptual scheme. Each new experience, finding or idea anywhere in science poses a potential challenge to these shared conceptual understandings.

A new discovery is frequently as difficult to describe as a sunset. In at-

---

1. This percentage is based on Derek Price's estimate that 80 to 90 percent of all scientists who have ever lived are alive today (1963:1) and the OECD's (1968) estimate that one-half of all scientists alive today are in the United States. Together these indicate that between 40 and 45 percent of all scientists who have ever lived presently reside in the United States.

tempting to state his discovery in terms understandable by interested scientists, the discoverer tries to fit it into his conceptual scheme. If it does not fit, he tries to expand or change that scheme so that the discovery does fit. Each alteration must be shared with all other interested scientists so that its content and form can be modified by subsequent discussion and critique.

Scientists have developed a series of communication tools ranging from mathematical formulas to diagrams and graphs which enable the "shorthand" expression of ideas. Learning to read those expressions and ideas is part of a scientist's training. Those trying to understand science without some of this training will have as much difficulty as if they were in Germany with no German language background. Obviously, then, nonscientists have a very low level of actual understanding of science. C. P. Snow (1959) observed that even educated men are scientifically illiterate if they have not studied science.

The history and philosophy of science have long been established areas of interest in science (Hall, 1963; Kuhn, 1970). The sociology of science, however, is a relatively recent specialty of sociology. Robert Merton's study of the rise of science in England was an early work in that area, and his papers since have been important landmarks in the study of science (Merton, 1938, 1959, 1961). The sociology of science reader by Barber and Hirsch (1962) and Barber's *Science and the Social Order* (1966) mark points at which the sociology of science began to be recognized as an important specialty. Warren Hagstrom (1965), Norman Storer (1966), and Joseph Ben-David (1971) are sociologists who have written major monographs on science.

The following pages will introduce the liberal arts student—and perhaps some scientists also—to the scientist's social world. We will examine four topics:
   a. Who and what are the scientists?
   b. What is a scientific career?
   c. What is the social organization of science?
   d. How does science affect society in general?

# I. Scientists: Occupations, Community, and Work Style

Scientists are persons of all ages, both sexes, all races, and many occupations who use specific skills and a specific language. Most of them have been trained by other scientists, generally within educational institutions. Educational credentials such as the Ph.D. are often used to differentiate scientists from each other as well as from nonscientists. When we talk about science, we must recognize the existence of two different structures: the scientific occupation and the scientific community (Hagstrom, 1965). We must also recognize that scientists have a particular work style.

# SCIENTIFIC OCCUPATIONS

The National Science Foundation long conducted biennial surveys which counted the number of scientists engaged in each officially determined scientific occupation.[2] Table 1 reports selected results from the 1970 survey.

### Table 1. Number of Scientists by Field and Highest Degree, 1970

| | Total | Highest Degree | | |
| | | Doctorate Number % | Master's Number % | Bachelor's Number % |
| --- | --- | --- | --- | --- |
| All Fields | 312,644 | 125,234(40)* | 93,540(30) | 83,740(27) |
| Chemistry | 86,980 | 29,985(34) | 19,178(22) | 36,514(27) |
| Earth and Marine Sciences | 23,756 | 5,555(23) | 8,006(34) | 9,910(42) |
| Atmospheric and Space Sciences | 6,637 | 693(10) | 1,882(28) | 3,089(46) |
| Physics | 36,336 | 16,631(46) | 12,215(34) | 7,392(20) |
| Mathematics | 24,400 | 8,033(33) | 11,975(49) | 4,349(18) |
| Computer Science | 11,324 | 1,010 (9) | 4,859(43) | 5,434(48) |
| Agricultural Science | 15,730 | 2,944(19) | 4,176(26) | 8,520(54) |
| Biological Science | 47,493 | 24,334(51) | 10,303(22) | 5,719(12) |
| Psychology | 26,271 | 17,593(67) | 8,433(32) | 215 (1) |
| Statistics | 2,953 | 1,157(39) | 1,256(42) | 532(18) |
| Economics | 13,386 | 7,225(54) | 5,008(37) | 1,082 (8) |
| Sociology | 7,658 | 3,690(48) | 3,181(42) | 758(10) |
| Political Science | 6,493 | 3,990(61) | 2,447(38) | 48 (1) |
| Anthropology | 1,325 | 1,260(95) | 36 (3) | 17 (1) |
| Linguistics | 1,902 | 1,134(60) | 585(31) | 161 (8) |

*Source:* National Science Foundation, 1971, pp. 13, 15, 17.

* Percentages do not add up to 100 because there are some scientists with professional medical degrees, without a bachelor's degree, or not reporting.

Note that roughly 40 percent of those in scientific occupations (125,234/ 312,644) have attained the Ph.D. (Doctor of Philosophy, highest earned degree in most fields). For specific fields of science, this percentage can be much higher or lower than for science as a whole. Most computer scientists (91 percent), for example, are not Ph.D.'s; but most anthropologists (95 percent) are.

The number of persons working in occupations clearly associated with science is much larger than the number of actual scientists. The Organisa-

2. The NSF approached scientists through professional organizations. Therefore, it tended naturally to overselect scientists who do research and depend on professional associations for publications while underselecting those who neither follow the literature closely nor publish in it.

tion of Economic and Community Development (OECD) estimated that 4.7 percent (4 million people in 1968) of the total work force were in science-related occupations. Of these, however, only 750,000 were actually scientists and 153,000, Ph.D.'s.[3] The bulk of the nonscientist population was composed of engineers (1.4 million, of whom only 17,000 were Ph.D.'s) and technicians of various varieties (1.6 million).

## Table 2. Scientific Personnel In United States, 1970 (est.)

| | |
|---|---|
| U.S. Population | 209.0 million |
| Work Force | 86.0 million |
| Total Scientific Population | 4.0 million (4.7% of work force) |
| Scientists | .74 million (153,000 Ph.D.'s) |
| Engineers | 1.4 million (17,000 Ph.D.'s) |
| Technicians | 1.6 million (17,000 Ph.D.'s) |
| Science Teachers | .3 million |

Source: OECD, Reviews of National Science Policy: United States, 1968, p. 45.

Whatever the counting method used, it is evident that the number of persons engaged in scientific occupations has steadily increased at a rate which, if maintained, would make every American man, woman, cat, and dog a scientist by 1990 (Martino, 1969).

## Table 3. U.S. Population and Population in College

| | U.S. Population (millions) | Number Enrolled in College (millions) | Percentage of 18–21 enrolled | Completed College (millions) |
|---|---|---|---|---|
| 1940 | 131.7 | 1.7 | 15.6 | 3.4 |
| 1950 | 150.7 | 2.5 | 29.6 | 5.2 |
| 1960 | 178.5 | 3.4 | 34.9 | 7.6 |
| 1970 (est) | 206.0 | — | 38.4 (1966) | |

Source: Statistical Abstract of U.S., 1969, p. 5.

Consider what this expansion rate suggests about who scientists are and how they fit into American society. The United States population has been increasing (see table 3), and we have been educating a larger percentage of

3. The OECD's larger estimate for the total number of scientists is partly due to the fact that it uses *estimates* of the employed population working as scientists; the NSF counts only those scientists responding to questionnaires.

persons in the prime college years (ages 18–21). These two effects have combined to increase the absolute number of persons in college. The process which produces scientists skims its trainees from the top groups (academically) of those who complete college (see pp. 14ff). Not only, then, is the potential supply of scientists up; so also is the supply of trained people in general. Table 4 shows that the number of people employed in jobs classified as professional and technical has been increasing over the last twenty years.

### Table 4. Persons in Work Force With Professional and Technical Training

|      | Millions | Percentage of Labor Force |
|------|----------|---------------------------|
| 1950 | 4.49     | N.A.                      |
| 1955 | 5.79     | N.A.                      |
| 1960 | 7.48     | 12.0                      |
| 1965 | 8.88     | 13.0                      |
| 1969 | 10.95    | 14.3                      |

*Source:* Compiled from OECD, 1968, pp. 42–45.

N.A. = not available.

As any economist will tell you, however, supply is only part of the picture. Until recently, demand has been increasing. The larger number of students in high schools and colleges has demanded more teachers who are now generally required to be Ph.D.'s. Of all the scientists employed by educational institutions, 58 percent have the Ph.D. degree. The increased federal funding of research-heavy areas such as space and weapons systems development brought a concomitant increase in the number of persons employed in these programs. While 1969–1972 saw only small increases or even declines in these expenditures, the need to spread identical amounts of money over an expanding number of personnel gave the impression of a very great loss. The demand for scientists who have completed college (table 3, col. 4) and for technicians and professionals has managed, through 1970, to provide employment for most, even though many have been unable to obtain research funds.

It was unlikely, however, that the 1960s growth rate for scientific occupations could continue. The demand side of the market for scientific talent has become more selective in the 1970s, and the supply side will occasionally err in the kinds and numbers of scientific graduates offered. If we consider that most Ph.D.'s graduating in 1972 began their graduate school specialty in 1968 when the demand picture was very different (e.g., the space program was still expanding quite rapidly), it is not surprising that some Ph.D.'s began their post-doctoral careers as cab or truck drivers. (See p. 16.)

# SCIENTIFIC COMMUNITY

As was noted earlier, the actual scientific community is smaller than the number of people with scientific occupations. The scientific community, strictly speaking, includes groups of scientists who are actively involved in doing research, publishing their results, and communicating with one another. This community's small size is indicated by the fact that many scientists with Ph.D. degrees never publish more than one paper, usually a part of their doctoral dissertation, and that half of all Ph.D.'s never publish at all (D. deS. Price, 1963: 40–45).[4]

The active scientific community, then, may be as small as 25 percent of the total population of scientists, perhaps 180,000 of the 750,000 estimated by the OECD. The active groups cluster at major institutions for access to facilities, prestige, intellectual stimulation, publication, and other advantages. These institutions also weed out nonactive members. The community is not only smaller than the total population of scientists; it is drawn from the top of that population with respect to training (including a much higher percentage of Ph.D.'s than the total science population) and status. For any given discipline, active scientists constitute its core, providing both impact on that field's content and status among those institutions which employ them. More broadly, that active core usually includes scientists who are central to the communication network which encompasses all of science.

The scientist who wants evaluation of his results, information from someone he trusts, or recognition of his work from those who do work similar to his, needs very detailed information about the hundreds or even thousands of persons who say that they share an occupational specialty with him. One way to obtain this information is through the loose network of scientists who communicate with each other about their research. For further discussion of this network, see pages 25–27.

# THE SCIENTIST'S WORK STYLE

It is hazardous to generalize about the scientific work style since it can be somewhat affected by occupational location. However, some impressions can be gained from the NSF data in table 5 which describes the primary occupations of most scientists as research, management or administration, teaching, production and inspection, consulting and exploration, forecasting and reporting, and design and development. (The numbers in design and development can be discovered from table 5 by adding the basic and applied research columns together and subtracting the result from the "R & D" row.) In 1970, about 42 percent of all scientists responding to the NSF

---

4. The difficulty with Price's estimate, which refers to the discipline of chemistry, is that he based it only on "quality" journals. He did not determine whether or how often scientists publish in journals at other levels of significance. Even given this qualification, though, the startling fact still remains that most scientists publish very little. Similar data have been found for all other disciplines that have been studied.

# Table 5. Scientific Work By Employer Categories

| | Education Number (%) | Industry Number (%) | Federal Government Number (%) |
|---|---|---|---|
| Research and Development | 39,843(30)* | 33,803(35) | 12,860(41) |
| Basic | 28,764(22) | 5,655 (6) | 5,538(18) |
| Applied | 10,543 (8) | 18,155(19) | 6,643(21) |
| Development | 536(—) | 9,993(10) | 689 (2) |
| Manager or Administration | 11,938 (9) | 33,321(34) | 11,417(37) |
| of (R and D) | 3,997 (3) | 16,779(17) | 4,794(15) |
| Teaching | 70,302(54) | 274(—) | 355 (1) |
| Production and Inspection | 203(—) | 13,437(14) | 952 (3) |
| Consulting | 2,672 (2) | 4,258 (4) | 968 (3) |
| Exploration, Forecasting, Reporting | 840 (1) | 7,783 (8) | 2,592 (8) |
| Other** | 4,591 (3) | 4,666 (5) | 1,974 (6) |
| Total (% of 312,644) | 130,389(42) | 97,542(31) | 31,118(10) |

*Source:* National Science Foundation, *American Science Manpower,* 1972, pp. 19, 22, 24.

\* Percentage of column total.

\*\* Includes no report.

questionnaire worked in educational institutions, 31 percent in industry and business, and 10 percent in the federal government (NSF, 1971:5).

Table 5 also shows that the scientist's primary kind of work depends on the type of employer. Roughly a third of all scientists in both education and industry do research. Three-quarters of the education group do basic research (the scientist generally sets his own problems) while one-quarter does applied research (in which problems are set by the desire for profit, prestige, academic advancement, or other extra-scientific goals). In industry, by contrast, only one in five researchers does basic research; the other four do applied research and development. Roughly one-third of the scientists employed by industry or government are involved in production, inspection, forecasting, exploration, and reporting.

A second major difference between industry and education lies in the degree of supervision done by scientists. Only 9 percent of all scientists in educational institutions manage others as a primary activity while almost a third of all industrial scientists do. A further difference is that 51 percent of the scientists in educational institutions teach as their primary activity.

Educational institutions (see table 6) are the largest employers of scientists, in part because of the system developed during and after World War II through which federal research grants and contracts were given to non-profit institutions, particularly universities and colleges. The second largest employers are industry and business; the federal government is a poor third; and other institutions employ very few scientists. As was noted earlier with respect to percentage of Ph.D.'s, however (see table 1), we can find strong differences among fields; for example, 66 percent of all computer

## Table 6. Number of Scientists, by Field and Type of Employer, 1970

| Field | Total | Educational Institutions Number (%) | Industry Number (%) | Federal Government Number (%) |
|---|---|---|---|---|
| All Fields | 312,644 | 130,389(42) | 97,542(31) | 31,118(10)* |
| Chemistry | 86,980 | 19,844(23) | 50,980(58) | 5,029(6)** |
| Earth and Marine Sciences | 23,756 | 6,535(28) | 9,973(42) | 3,046(13) |
| Atmospheric and Space Sciences | 6,637 | 1,017(15) | 768(12) | 2,053(31) |
| Physics | 36,336 | 18,085(50) | 9,819(27) | 3,912(11) |
| Mathematics | 24,400 | 14,020(57) | 6,364(26) | 1,243 (5) |
| Computer Sciences | 11,324 | 1,603(14) | 7,462(66) | 882 (8) |
| Agricultural Sciences | 15,730 | 3,563(23) | 2,581(16) | 5,725(36) |
| Biological Sciences | 47,493 | 28,778(60) | 4,836(10) | 4,912(10) |
| Psychology | 26,271 | 14,865(56) | 1,943 (7) | 1,549 (6) |
| Statistics | 2,953 | 1,075(36) | 773(26) | 599(20) |
| Economics | 13,386 | 7,845(59) | 1,822(14) | 1,513(11) |
| Sociology | 7,658 | 5,675(74) | 129 (2) | 224 (3) |
| Political Science | 6,493 | 4,998(77) | 117 (2) | 349 (5) |
| Anthropology | 1,325 | 1,070(81) | 11 (1) | 29 (2) |
| Linguistics | 1,902 | 1,417(74) | 54 (3) | 53 (3) |

*Source:* National Science Foundation, 1971, pp. 13, 15, 17.

\* Percentage of total in field.

\*\* Percentages do not add to 100 because there are other employers and unemployed. Unemployment is at 6% for all scientists.

scientists are employed by industry whereas 81 percent of all anthropologists work in educational institutions.

Scientists may have freedom of research choice in industrial and other noneducational laboratories, but such freedom is not common. Bell Laboratories and Du Pont's Central Research Laboratory, for example, show a degree of freedom as great as or greater than that of any university, but they are rather unique (Cordtz, 1971). The control of research topics and processes can be more clearly seen in an industrial laboratory, but control can be exerted by a university through budget constraints and requests by a laboratory director that scientists work on specific projects. Every laboratory finds some way to balance individual initiative and management control. It is likely that a "star" performer, wherever he is, will be encouraged to go his own way. (Research professors, usually the stars in the university system, and adjunct professors may teach very little or not at all.) A marginal performer in the university system is likely to have job security via the tenure system, and as long as his research does not draw too heavily on the institution's resources, he will be permitted to continue.

A scientist can be a researcher in almost any occupational position and live the research work style *as long as he is successful.* But each institution may judge success a little differently. For example, most industrial laboratories want some profitable return from research, and most government units have specific program goals. A less successful researcher in one of these situations may find his work becoming more administrative and managerial. In an educational institution, he may find himself carrying a heavier teaching load than his colleague who does very well with his research. (See p. 12 for other criteria of success.)

A research scientist's daily life usually includes reading articles and books, thinking, writing up results, manipulating apparatuses, talking with colleagues, administering funds and personnel, attending meetings, and (at a university) teaching and advising students.

A report from the American Psychological Association (1963:15–16) on the activities of 78 psychologists emphasized the importance which scientists place on communication:

> *The picture one receives after surveying these seventy-eight accounts of two weeks of scientific activity, recorded in seventy-eight different ways . . . is on the whole an encouraging one. . . . Possibly this group is an exceptional group, for, busy as they apparently were, the willingness with which they made every effort to supply information and to do so immediately was generally evident. Equally obvious was their readiness to talk at any time during a crowded day with students or colleagues needing information or assistance . . . it would appear that such interchange was usual and a result of their intense interest in, and preoccupation with, their research and with the general subject matter of psychology.*

There are certainly individual differences in how scientists approach their various activities, but the basic activities remain the same.

The process of doing research in any environment involves (1) setting the question which is to be answered by research (the question can be general or specifically related to someone's on-going work), and (2) the development of facilities to answer this question. This latter typically includes acquiring some technical apparatus although most of the equipment needed is usually present.[5]

Large questions subsume smaller problems. Solving small problems is likely to require development of a subsidiary method which is necessary before the chosen question can be approached with the chosen technique (e.g., slide-staining technique must precede microscopy; at a more technical level, those techniques of protein chemistry which permit replacement of certain atoms in proteins must be perfected before X-ray crystallography can be done). Such smaller questions may be assigned to another laboratory either through formal subcontracting or collaboration. Sometimes a specially trained person is hired if a great deal of special work must be done, and his presence would not require radical shifting of the laboratory's

---

5. The general pattern is that a scientist takes problems within an area that uses equipment already available to him and has done so as a matter of economy rather than of scientific interest. Thus the apparatus is an important factor in the active selection of problems. Only a favored few have managed significant changes in the material and personnel characteristics of their laboratories.

equipment. Or some regular, informal process of cooperation may exist (e.g., that between protein chemists and protein synthesizers).

The type of communication which brings together research scientists and technicians does not necessarily occur within the boundaries of the communication network mentioned earlier. However, when a scientist becomes expert at a technique which is secondary to his initial research question and then continues to use that technique in later work, his own or that of others, he may provide the spark which begins a new network or expands an older one. For example, the application of physical science techniques to biological science problems, beginning in the late 1930s, resulted in the eventual development of molecular biology (Cairns, Stent, and Watson, 1966; Mullins, 1972).

When methods are applied to questions, data result. The important task then becomes interpretation of this data. Some data will be insignificant and never reported. Some will become part of a field's "oral tradition" and never be published (e.g., the product which results from mixing compounds X and Y makes the flask too hot to handle; mix them on a bench!). Some data will be considered significant enough to be published, usually with some interpretation.

The usual research paper, which is a scientist's means of formal social communication with his occupational peers, has three parts. First the research must be set in context. The scientist must describe previous work, citing results and stating the new question his work has answered. Second, he describes the method and materials used. The new material follows and may constitute as little as one-tenth of the paper's bulk. Finally he interprets this new data and shows how it is related to the total knowledge of his field. The excitement and/or tension generated by this juxtaposition of ideas with other ideas and results is sufficient to motivate the creative scientist (Koestler, 1964). Active scientists seek such tension as do other creative people. To paraphrase a famous scientist, "If you haven't the talent to be an artist, what is there to be except a scientist?"[6]

Creative scientific work and the acceptable communication of such work often make scientists seem like jungle prospectors who, having been airdropped into a trackless waste, have cleared out the underbrush and worked their way back to civilization. The scientist's jungle is created by his imagination. As he answers questions, he slowly works his way back to that group of other scientists who stand outside his jungle. The highest rewards (such as Nobel Prizes, research money, jobs) are given to those who come back with (1) the clearest directions to the place they've been, and (2) a good reason why others would want to go there. There are no guidelines to the possible success or failure of a proposed project, only a hunch which makes a scientist place a bet (his future) on a specific alternative. This analogy can be stretched a bit further: those who go short distances and return frequently receive low rewards; those who go farther

---

6. Max Delbrück, quoted by N. Visconti, "Mating Theory," in Cairns, Stent, and Watson (1966).

receive greater rewards. Scientists often run into half-cut "trails" as they work their ways from one "place" to another. These trails are the results left by forgotten pioneers who never returned to the main clearing (e.g., Mendel's work on genetics, which lay undiscovered for fifty years). Some trails leading away from a clearing are never retraveled; these are the products of old research results whose discoverers were never able to convince anyone that a particular piece of work was worth doing (e.g., invariant theory in mathematics; see Fisher, 1967:216–244).

The creative research activity just described probably constitutes every active scientist's ideal occupation. Further, the research work style is the only uniquely scientific work style. All other activities related to science (management, teaching, production, and so on) resemble the activities of nonresearchers in scientific areas. The researcher may thus possess skills that nonresearchers lack, but if he is not utilizing those skills, his work style will not differ significantly from that of the nonresearchers with whom he works.

# II. The Scientific Career

## THE BEGINNING

Most scientists begin as winners in the academic competition. They have always done well in school. As soon as they showed competence, for example in mathematics, they were pointed toward a career in science, engineering or some technical area. The typical, successful training of a scientist is outlined in table 7. The numbers placed along the left side of

### Table 7. Selection of Potential Scientists

| Age | School | Characteristic |
| --- | --- | --- |
| 6-13 | Grade school | Mathematics skills recognized |
| 14-18 | High school | College prep. diploma; science track |
| 19-20 | College, years 1 and 2 | Begins major study in mathematics and science |
| 21-22 | College, years 3 and 4 | Science major |
| 22-24 | Graduate school: early | Picks field |
| 24-26 | Graduate school: finishing years | Picks specialty and thesis topic |
| 26-28 (Approx.) | First job | Selects career style |

Source: Cooley, 1958.

the table indicate average ages. They show that the new professional is generally between 26 and 28 years old and has been in school 20 to 22 years. As the table shows, selection for mathematical ability and scientific interest occurs at every level of the educational process.

One study suggests that junior high school (grades 7–9) is the last period in which significant numbers of students enter the Potential Scientists Pool (PSP; see Cooley, 1958). Many studies have shown that post-high school transfers between the PSP and nonscientist groups are one-way (from the PSP to the nonscience). Persons who do not begin a science curriculum during high school are rarely found later in college or graduate school science programs. Some persons with ability are lost along the way through lack of money, motivation, or information about the system. Selection at the high school level removes fifty percent of the population and only thirteen percent of those who finish college go on to graduate school (OECD, 1968: 56). Of those who begin graduate school in the sciences, only six percent complete their degrees (OECD, 1968:52, 56).

Technical jobs requiring different amounts of training are available, and a person leaving the educational process can generally find some job available which requires his amount of education and no more. For example, a person with two years of college is eligible for a job requiring a high school diploma and some college. If he went to junior college, he may have an associate degree and special training which opens other opportunities. Such jobs are often available in teaching, technical aspects of research operations, technical sales, etc.

One difficulty with the system as presently organized is that those pursuing a Ph.D. degree must often take full-time jobs before finishing their theses, thus raising the possibility that they will never complete the degree requirements. One study showed that only one in every five candidates (21.6 percent) was able to finish his doctorate without taking an outside job (Wilson, 1965). The necessity for taking a nonthesis related job obviously lengthens the time needed to finish Ph.D. training. The study also showed that the graduate student career averaged 4.2 years, with 25 percent of the respondents taking longer than five years.

The Ph.D. degree is a "union card" signifying that its holder has "arrived." It does not necessarily mark the absolute end of the educational path, but it frequently does. First publication marks a scientist's final "coming of age." Given the publication figures cited earlier, most scientists remain in a perpetual adolescence, never publishing anything beyond their theses and remaining generally uninvolved in the scientific community and its communication network.

Post-doctoral fellowships in the natural sciences are available in many countries for both new and older Ph.D.'s, whether native to a country or foreign. The post-doctoral student is a professional with some independence, but, just as in graduate training, he is under the supervision of an older scientist who, usually, is someone other than his graduate work supervisor. Thus, as he moves into his chosen field, he gains a second sponsor and more advanced skills.

The relationship between student and teacher-sponsor in graduate study is very important to a scientist's future career. He is channeled into his first scientific problem by his professors who teach him the particular technical capabilities, concepts, approaches, and commitment that will later characterize his work. (Most scientists do not, during their careers, significantly alter the ideas they receive while in graduate school; Kuhn, 1970.) The sponsorship aspect of the teacher-student relationship is also important. Recommendations by a teacher of his former students are crucial to decisions both to hire them initially and, subsequently, to grant them academic tenure. The graduate sponsor can be particularly important in helping him acquire a good first job.

The importance of proper graduate training under a prominent sponsor cannot be overemphasized. It has been demonstrated that highly productive people, particularly those who have received major recognition, tend to be the students of highly productive men who have themselves received major recognition (Zuckerman, 1967). Even for the less productive scientist, proper training is important to a career. The more productive scientists tend to cluster at major universities (Cole and Cole, 1967), and the major universities turn out the majority of new professionals (National Science Foundation, 1969). We know from experience that top students are attracted to major universities partly by counseling, partly by the availability of research resources, and partly by prestige. We also know that within major universities the difference in quality between top faculty members and the rest is not as great as in other institutions. All of these factors work together to promote and enhance an atmosphere of excellence (Hagstrom, 1968). This total system acts to bring groups of good students under the tutelage of good faculties.

It is possible, however, that the graduate education system may waste much of this advantage (revision of graduate education has been a major topic at most institutions). The use of graduate students as cheap labor on projects that do not advance their education, the indifference of faculty to student problems or desires, and the lack of good (i.e., productive for the students) student-faculty contacts have become major concerns.

A new element in the graduate education picture has been the protest and turmoil on college campuses. The large universities which have been the centers of graduate excellence (Wisconsin, Berkeley, Columbia, Harvard, Michigan, Chicago, etc.) became, in the late '60s, centers for tension, protest, and even violence. Part of the diminishing financial support for these institutions (see pages 27–33 on financing) can be seen as a reaction to these difficulties. Part, however, results from the relation of science to the general society.

# CAREER CONTINUATION

Being trained and obtaining status as a new professional constitute only the beginning of a career. Subsequently, most scientists' professional careers are unproductive and unnoticed by most of society. They may be good teachers, advisers, citizens, and employees, but they are not productive scientists. Productive members of the scientific community are rare. Publication (frequently used to measure productivity) for a given group follows Lotka's "law of distribution" which states, in part, that five percent of the scientists will produce fifty percent of the literature in a field, while fifty percent will produce only five percent (D. deS. Price, 1963:40–50).

It is possible for a scientist to gain rewards as a member of either an organization, the scientific community, or both. A member of an organization joins committees, works to improve his institution and his own standing in it, and generally focuses his attention locally. A member of the scientific community spends most of his time on his professional associations (Glaser, 1964).

Two facts of scientific life lead to individual tragedy. First, scientists "age"; most do not change their techniques or ideas after their Ph.D. training has been completed. Second, new ideas and techniques develop with some regularity. In engineering, where skills are directly usable in technological processes, it can become painfully obvious very rapidly when a person no longer has the skills to do a particular job. Engineers call this phenomenon "technological obsolescence," and some effort has been made to provide retraining. Scientific obsolescence is less a matter of technology than of conceptual schemes. Scientists who simply fail to understand or to accept changes in the way in which a field's conceptual scheme is being modified gradually become uninvolved in the current ideas of their disciplines. They may be excellent administrators, teachers, or deans, but they are no longer productive scientists.

Unemployment has recently become a specter for scientists. A combination of too many scientists (in certain fields) and the reduction, during the late '60s and early '70s, of general funding support for science has produced a situation in which some scientists are unemployed. Less than one percent (1.2/297) of a sample of scientists in general reported themselves as unemployed in 1968. The percentage for physics in 1970, however, was closer to four percent (*Newsweek,* March, 1970). To be sure, most fields showed figures closer to full employment or at least lower unemployment figures than physics. Nevertheless, some individual cases interviewed by *Newsweek* were rather alarming—e.g., the Ph.D. in particle physics who was driving a cab in New York. Part of this unemployment will be reduced by using Ph.D.'s in positions formerly requiring master's or bachelor's degrees, but this compensation will only shift the unemployment problem to a different educational level. The secondary school teaching profession was itself overstaffed in 1970 for the first time in many years, and future prospects are for continuation of this situation (*Newsweek,* June, 1970).

Unemployment is particularly critical for scientists because of the knowledge obsolescence discussed above. In many fast-growing areas, if a scientist has not kept abreast of current research with informal contacts available through his employment, he may be as much as two years outdated even if he has a new Ph.D. degree.

The scientist makes a major choice when he takes his first job. If he goes into industry or government, his chances for subsequent academic employment, with its higher prestige, are very small. If he remains within the academic world, however, he can expect little difficulty in shifting to industrial or government employment later should he so desire. Many scientists have moved from academic to partial employment in industrial areas as a result of inventions developed from their academic work. The early 1960s marked a high point of such activity. However, the general slowdown in technologically oriented small businesses in the late 1960s and early 1970s is likely to retard such occurrences.

The highly successful academic scientist generally becomes an entrepreneur with a laboratory of several other scientists, post-doctoral students, graduate students, and so on. He will frequently move into a position that involves much more administration than research.[7]

Scientists who want to remain within the scientific community and within the group of researchers active in basic research problems will stay employed either in a university setting or in an exceptional outside laboratory. There they are best able to communicate with other scientists, publish, and work on problems of interest to them and to the scientific community.

# III. Social Organization

## THE STRUCTURE OF SCIENCE

Social organization is important to scientific study because social groupings influence a scientist's research decisions at crucial points. Some of these groupings are formal organizations like laboratory research groups and disciplines. Others are less formal but perhaps are more important—e.g., groups of scientists who communicate with one another about their research or groups of graduate students who worked in the same laboratory at the same time and have continued close contact after receiving degrees.

Research, the primary output of the scientific community, is done according to a series of related, and only partly objective, rules. These rules are especially effective when scientists feel threatened by a lack of knowledge. Each scientist's decisions are affected by the rules, mythology, and decisions of the groups to which he belongs. In purely rational decision-making terms, a scientist gains others' support for the decisions he is making so that he can "cut his losses" if he is eventually proven wrong; if he is proven right, he gains not only research credit but the appreciation of others and thus increases his gains. The scientist who works in this way is clearly operating from a position of strength.

7. Further study may eventually demonstrate that the movement of research scientists into administrative positions provides examples of the Peter Principle at work. See Peter and Hull (1969).

The group gains influence in several ways. Tough initiation rites result in strong attachments; they make the group more attractive (Eisenstadt, 1964: 62). The more attraction exists, the more conformity appears (Hare, 1962: 370). Even technical groups, if they are successful, become friendship groups (Bennis, 1956). But they also become larger as their success attracts new members (Mullins, 1972) and subsequently split and reform (Hagstrom, 1965).

The scientific community's structure cannot be depicted on an organization chart; the structure resembles that of a primitive tribe. We have already mentioned some similar elements between the two:

1. most of the population is young and the birth rate is high (see p. 6).
2. a few central figures do most of the work (see p. 8).
3. a long arduous initiation provides motivation (see above).

Others of these surface characteristics are:

4. a rich symbolic life expressed socially and a rich social life expressed symbolically. Clans and totems symbolize these aspects in scientific life as they do in tribal life;
5. a barter economy;
6. population stratified into age sets;
7. a simple, two-level polity.

The symbolic life of science is expressed in three ways. One is the naming of specialties, disciplines, meetings, and other collectivities of scientists. This naming resembles a primitive tribe's association with a totem which supposedly has some power. A field of study, be it a discipline or a specialty, has a distinct name (e.g., geology) and scientists who practice it (e.g., geologists). It has a graduate training program to "give birth" to members, textbooks primarily about the area, and departments or institutes baptized with the name of the field (e.g., Department of Geology). It may be part of another area (e.g., geology has become part of earth sciences, which also include meteorology, oceanography, etc.), but it must have the minimum elements just noted to be considered a field (See pp. 25–26 for discussion of how a new field gets started).

Symbolic life is also expressed in the elaborate and detailed nature of scientific knowledge itself. Any textbook about a science will demonstrate this complexity. The third expression is in myths and stories. Scientists often tell stories about one another which may begin, for example, "Did you hear about the explosion in old Kunkel's lab?" Many scientists use cautionary tales, just-so stories, and other elements of myths to make points and to provide legitimacy (e.g., "As Pasteur said . . . "). A scientist's whole work life, then, is permeated by symbols.

The barter economy is not immediately visible. A scientist uses information as the "money" in his economy. He trades information (with other scientists) for other information, never for money and rarely directly for prestige (Hagstrom, 1965).

Scientists can always be classified into one or another age set. When a new scientist is a student or junior member of a laboratory, he is usually just one of several people whose status is similar to his. These scientists belong to the same age set. One study of research scientists (biochemists) (Mullins, 1966) showed that every scientist reported keeping close contact with those of his fellow graduate students who had been in his same laboratory. Some kept contact with even larger numbers from graduate training days even though the persons involved had moved into different specialties within biochemistry, or even into different science areas altogether (there was some tendency to drop those who had moved out of science entirely). Scientists having such relationships saw each other an average of once a year, and many letters were exchanged among them. Sociologists should not be at all surprised at this development of close ties among those who pass initiation rites together.

The same study showed that many scientists had formed similarly close ties with those passing through graduate school rites a short time before or after them. Also during graduate school, a budding scientist usually established relations with one or two of the senior people in his laboratory. As such units were broken up locally by the gradual receipt of degrees, the new Ph.D.'s tended to form friendships, and often working partnerships, with other young Ph.D.'s. Truly collaborative research, it should be noted, tends to be done only by young people.

As scientists grow older, their research group relationships tend more and more to be on a senior-junior basis; whatever close, work-oriented friendships exist in a research group are generally among students and young Ph.D.'s. A similar friendship pattern exists among the "old boys" of a field, but they rarely work in the same laboratory. It is normally between well-known and successful scientists that this pattern is noticed, but it exists among the less successful as well. Such scientific friends can often do favors for each other. Conversely, enemies can be influential in blocking paths to publication, research funds, etc. Snow (1961) notes that in prewar England, where the entire scientific community was rather small, a small group of peers (in age, not nobility) was able to act as a board of governors for science.

To the anthropologically oriented, these age strata look like the age sets among the Neur (they also exist in many other tribes) described by Evans-Pritchard (1940) in which the initiation class forms an important basis for social organization. The phenomena of one-to-one (but with parents instead of sponsors) and one-to-many relations (but with social rather than graduate school peers) can be observed also in the American youth culture; there also it establishes social grouping by age.

This comparison is not as far-fetched as it might seem. The student is a child in science. He must be trained and taught the values and norms of scientific behavior. For years the master-student relation of ancient scholarship has been the mechanism for teaching. The teacher or master may be any older or more experienced scientist with whom the student has contact. Usually this system is institutionalized in the structure of the educational system by an office such as advisor or committee chairman. The other members of a laboratory or department will have students of their own and will limit their time with any student of another professor.

Fellow students tend to help each other in their work, talk frequently in bull sessions, and share similar problems. Because social relationships frequently revolve around work relations, graduate students often play and party together. There is even some tendency for these social arrangements to extend into marriage (a sample of biochemists showed that 4 of 45 sample members were married to other sample members; Mullins, 1966).

The above discussion seems to indicate not sets of people divided by age but rather a continuing stream, bound in continuity from the oldest graduate of an institution to the youngest and bridged by the sponsor-student tie. Some of this stream does exist, but most of these groups are gradually dissolved over time by career patterns such as changes in the personnel of departments; the failure of most students to become produc-tive members of their disciplines; and developments affecting job avail-ability, research support, and research luck.

Finally, scientific organization has a simple, two-level polity: a few top scientists and everybody else. These top scientists, particularly in the United States, sit on tenure committees, fund-granting advisory boards, journal editorial boards, and so forth. These positions give their holders consider-able authority within science (D. K. Price, 1965).

The social organization of science is clearly complex, but we can find some fundamental aspects to its structure. We have already discussed one important factor, a scientist's training and its effect on his subsequent career. A second factor is most important as it relates to the scientific occupation. It stems from the fact that scientists are rarely self-employed. Science itself earns no money; it has to be justified as an economic, political, or educational enterprise. Institutions, most of them local, can thus become the glue that holds together a given location's group of scientists. A third factor is communication, which is aided or directed by one or another kind of formal association.

The fourth fundamental of scientific organization is change. No theory is final, no finding permanent, and no organization lasts very long. (Even seeming exceptions, e.g., England's Royal Society, founded in the seven-teenth century, have undergone massive changes in form.) The whole of modern science began just over 300 years ago, and natural philosophy cov-ered the entire area for half of that time. Most areas of science today are younger than professional baseball. Many of these areas (e.g., molecular biology), although their practitioners can find an occasional interesting aside in Aristotle or some other long-dead source, are younger than most college students.

# LOCAL ORGANIZATION

A scientist usually belongs to several social organizations made up of colleagues and assistants in his work location. Even the minority of scien-tists who work largely alone are likely to be part of an organization devoted to research. That organization may be an industrial research operation, a governmental facility, or a university laboratory, but it always has two characteristics: it is a formal organization with a budget, director, and staff; and it usually includes several research groups which work on specific problems or problem areas. Each research group usually has a leader and at least one technician. If a local group is part of a larger organization such as a university, the activity and structure of that group depend on the availability of money either in grant or budget form and the preferences and rules set by the host organization that is paying the bills.

Such a group is very well suited to carrying out normal scientific research projects. It can ask for funds, carry out research, write up results, and, while the first research is coming to an end, request money for another project. Most of the studies on the character of local research groups are largely inconclusive. Pelz and Andrews (1966:7) found that "in effective older groups, the members interacted vigorously and preferred each other as collaborators, yet they held each other at an emotional distance and felt free to disagree on technical strategies. . . . The scientists and engineers whom we studied did effective work under conditions that were not completely comfortable but contained 'creative tensions' among forces pulling in different directions." Pelz and Andrews (1966:297–302) also found differences in publication rates (government, high; academics and industry, much lower) and motivation structures (industry rewarded profitable products; academic and government organizations rewarded publication).

## COMMUNICATION

The social structure of the scientific community (disciplines, meetings, specialties, networks, etc.) is organized around communication. Menzel (1962:419) defines scientific communication as:

> . . . the totality of publications, facilities, occasions, institutional arrangements, and customs which affect the direct or indirect transmissions of scientific messages among scientists.

Scientific research is a highly uncertain undertaking, and regular scientific communication helps to reduce those uncertainties by:

1. Answering specific questions. If a scientist finds an answer that he needs in another scientist's work, he will probably search that scientist's other research for answers to other questions that he has.
2. Keeping the scientist abreast of new developments. Current knowledge is essential if a scientist is to do useful work.
3. Giving the scientist a sense of the major trends in his own field and of the relative importance of his own work. These trends are those changes that seem to be currently productive and likely to remain so.
4. Helping the scientist acquire an understanding of a new field. This "brushing up" asserts a scientist's right to join a research area, usually by an appeal to common training in the broader discipline of which that field is a part.
5. Verifying the reliability of information by additional testimony. Verifi-

cation requires (a) nonfalsification of a result by data (Popper, 1959), and (b) the building of a case that will convince a relevant body of skeptical others that a proposition is, indeed, true and useful (Ziman, 1968).

6. Redirecting and broadening a scientist's span of interest and attention.
7. Obtaining critical response and/or attaching a reward to a piece of work.[8]

Scientists, then, must communicate with each other, but this communication threatens to bury all of them under a mountainous pile of paper. The President's Science Advisory Council estimated that two million papers were published in 1968. There were also 100,000 informal governmental reports and at least that many informal reports elsewhere (U.S.P.S.A.C., 1968). Sorting through this information is a major problem which has been only partially solved by abstracting services, computerized reporting services, and specialized libraries.

In spite of this flow of paper, the most striking aspect of the scientific communication system is not the quantity of information but its redundancy. Each idea or piece of information usually appears in three different formats: oral presentation for meetings, conventions, etc.; internal technical report for the agency that supported the research; and journal article or monograph printed for the general scientific public. The first two formats reach a fairly specialized audience; the third, a more general one. This availability of more than one source provides the widest possible access for anyone who might want a particular piece of information; it also provides potential entrée to that network of people (see pages 25–27) who might know more about a particular subject.

Informal oral and written communication permits the exchange of ideas, results, and information for comment and evaluation—an exchange which may provide the primary reward for a scientist (Hagstrom, 1965). A scientist's informal communication of his initial ideas is the easiest form because those with whom he exchanges ideas usually share his background knowledge. As an idea is communicated and becomes invested with time and work, it becomes increasingly expensive to alter or replace. Publication requires even more time and effort because an author must communicate to a wider audience than usual, one which probably does not share his particular scientific background as closely as do his colleagues.

Oral communication occurs either at meetings or in ordinary conversation. The average number of meetings announced in *Science* (the national weekly magazine of science) is nine per week. These meetings range across such groups as the Informal International Great Lakes Study Group, The Congress on Environmental Health Problems, The American Society of Animal Science, International Congress of Histochemistry and Cyto-

8. With the exception of the two points referenced to other sources, this list derives from Menzel (1962).

chemistry, The Fourth European Congress of Cardiology, 2nd International Symposium of the Physiology of Digestion in the Ruminant, a joint meeting of twenty-four societies and associations in connection with the annual meeting of the American Institute of Biological Sciences, and the Conference on Education in the Nuclear Power Era. Some of these meetings are clearly informal, others quite well organized; some are international in scope, others regional or national. Some are meetings of specialists in one area; others involve many areas.

At all of these meetings it is interesting to observe the variety of groupings that develop. Some gatherings are directed toward the past. "All those who attended Cornell University between 1947–49 are invited to a breakfast to be held . . . " or "all students of Dr. ——— are invited to cocktails at six in. . . . " These are groups concerned with the past history of a discipline or the careers of its practitioners. They remind one of a college reunion, an American Legion Convention, or similar past-oriented gatherings. Attendees discuss the past and "where we have come since then."

A second kind of meeting is the "Sociology of Education Breakfast" or the "Neurophysiological Lunch," to take two recent examples. These meetings are present-oriented, directed to specific problems in the field. Future-oriented meetings also exist. "A ten-year program for UNESCO and the social sciences" and a breakfast to discuss the future of chemotherapy are both concerned with the future of a subdiscipline or problem area. Such meetings are usually marked by highly authoritative speakers rather than by the bright younger scientists usually seen at present-oriented meetings. The reason is quite evident. Since no good analytical tools yet exist for predicting the future, the acceptance of predictions must be based on the quality of a speaker's past credentials rather than on the quality of his present contributions or future promise.

Preprints, journal articles, reprints, reviews and textbooks all constitute printed communication. The preprint and the reprint are both copies of the journal presentation form which is the official way of adding to scientific literature. Journal articles are usually reviewed by other scientists to determine whether they indeed present material new to the field in question and thus merit publication. A journal's board of editors and staff of volunteer reviewers are of considerable importance to the process of legitimizing scientific communication. They have the power to grant or deny one of science's major rewards. Of course, the more prestigious the journal, the greater the power of its editors and reviewers.

Table 8 lists nine types of scientific communication with the times at which each occurs, the audience toward which it is directed, the author's relation to the audience, the type of reception the presentation should receive, the legitimacy of the communication for further use, and the accepted standards for determining a communication's worth. Further characteristics that distinguish kinds of communication are the frequency with which a given communicator has addressed a specific audience and the degree to which an audience and speaker share a theory about how their

## Table 8. Meetings and Publications

| | Invited conference | Preprints | National and International conference | Journal | Reprints | Journal reviews and grad seminars | Text and classroom |
|---|---|---|---|---|---|---|---|
| Time (Months before or after publication) | −12 | −12 | −6 to +6 | 0 | 0 | +18 | +36 |
| Audience | specialty | specialty | (limited by attendance) discipline (and specialty) | discipline (limited by reading habits) | specialty | discipline and related areas | discipline and students (educational world) |
| Author's Position | colleague | asks for comment review | stakes reputation on correctness | stakes reputation on correctness | semi-authority | summarized evaluations; place his contribution | authority |
| Reception | discussion with personal contact | varies; some reaction | polite discussion | formal debate in journal | acceptance or rejection | acceptance | |
| Legitimacy of Source | quotable if published (may be changed) | not quotable (doubtful legitimacy) | tied to journal | completely legitimate; final form | like "journal" | based on journal and later review | based on reviews |
| Standards Type and Rigidity | loose originality and aptness; clarity | loose; like "invited conference" | rigid 1. structure and tightness of argument 2. relevance 3. accuracy | rigid 1. structure and tightness of argument 2. accuracy | like "journal"; availability | clarity, coherence, completeness | moderate clarity, accuracy, completeness (up to false closure) |

portion of the universe operates (and, consequently, how much explanation has to be done before the "meat" of a presentation can be given). Each kind of meeting and publication is characterized by a different combination of these factors.

# DYNAMISM AND CHANGE: COMMUNICATION NETWORKS

The college student is at his institution for only four or five years. During that time the number and size of departments and programs probably change several times, although these changes are slow and not very visible to the student. The social structure of science responds to changes in scientific interest, societal support, and the differential success of different fields in attracting and holding students. The dynamism which facilitates these changes is provided by the ability of communication networks to solidify (or not) into parts of social institutions. The lines of communication in these networks are more or less formalized within clusters. A cluster is not usually a face-to-face group. An occasional meeting will bring most of a cluster's scientists together, but actual communications are generally carried on between pairs of members by letter, telephone, or face-to-face conversation.

If a cluster or one of its members is successful in answering at least some of the research questions around which it has gathered and if it can give the impression of succeeding faster than members of other clusters already established in the field, it will draw more members from several sources. Some will come from the general field; these usually constitute a small but important group (Cairns, Stent, and Watson, 1966).

The largest source of new members is students of older members. A scientist's students usually accept his perspective on his work. Eventually those students have students and the result is an explosive growth in the area's numbers. The development of phage work into molecular biology provides a case in point (Mullins, 1972). Likewise, a lack of students can doom a viewpoint or line of research. Invariant theory in mathematics (Fisher, 1967) provides the case in point. Whatever the source, a successful cluster generally increases its membership while unsuccessful ones break up. At any given point in time, most existing clusters will be increasing in size.

Clusters, ironically enough, destroy themselves with success. As the number of members increases, the more active members tend to dominate those occasions on which the cluster meets. This domination begins a tendency toward stratification. The cluster's procedures become more formalized, and the atmosphere for discussion becomes less intimate. The irony in this development is that an intimate and free-flowing atmosphere for discussion is at least partly responsible for a cluster's appeal and successfulness. When new ideas are discussed less, a certain rigidity appears. Subclusters begin to form within the old cluster. The subcluster formation bears a clear resemblance to the age sets discussed earlier, and these peer groups are frequently the source from which new clusters develop.[9]

But let's follow the original cluster. The membership continues to increase, and for a time the new arrangement consists of peer groups that tend to isolate themselves but still report back to the main cluster. The result is a large network of intensive interaction with smaller but even more intensive spots of interaction around peer groups. The large cluster tends to be maintained for reasons other than those which originally brought its members together. If it has been successful over a period of time, its members are usually beginning to benefit from the general reward system of science. Their names are known, the field is beginning to have offices for its members, publications begin referring to the cluster's work, grants for new research are being awarded on the basis of past successes. These rewards result in allocation problems within the cluster. Who should get these rewards? Recommendations, reports, and review become important means for allocating the rewards available to the larger cluster (Cole and Cole, 1967).

The increase in members accelerates (about five years after the original members acquire students), and a geometric progression begins to build up steam. But a cluster rarely has sufficient new ideas to support an increasing number of scientists. As rewards increase, then, the field's original research challenge becomes less important, and its greatest problem becomes the maintenance of gains and what Kuhn (1970) refers to as "normal science." The cluster is "playing out the deck," doing the more or less expected, the excitement gone from the original clash of ideas (Koestler, 1964). The risk is also gone. The former cluster has become institutionalized as a subfield or specialty of its major field.[10] Phage work's institutionalization during the early 1960s as a specialty (molecular biology) of biology exemplifies this process.[11]

Institutionalization has usually begun when resource allocation is determined more and more by a subgroup of the cluster's most important scientists, a group which is "first among equals" with a cluster's other subgroups. This group derives its status from its involvement with the original discoveries that led to the cluster's initial foundation.

The processes of group dissolution and reformation are ongoing. When a scientist no longer wants (or is unable) to keep up with work in his area, he withdraws from his cluster and ceases to contribute to scientific com-

9. The one-to-one student-to-professor relationships noted earlier could not constitute the basis for cluster development; there are too few of them. The one-to-many graduate education relations, however, are a basis for such sets, and they are large enough to permit clusters to develop from them.

10. A *field*, conceptually speaking, is a unified group of persons studying some subject matter. A *discipline* is a set of courses taught at a university, usually represented by a department.

11. Strictly speaking, molecular biology includes more scientific perspectives (e.g., x-ray crystallography) than phage work alone. In the interests of keeping this paper's examples simple, however, I have omitted these other perspectives since they are of interest only to their followers.

munication. The payoff of a given course of action is uncertain, research is risky and often unprofitable, and science is competitive. Factors favoring a dropout, however, are balanced by other facts of scientific life. Active scientists fear humiliation, but their nature pushes them to continue creating and understanding original research. They are unwilling to settle back into jobs which require additional nonresearch tasks, such as teaching or administration. These factors all lead them to continue active participation in scientific communication (Reif, 1961).

# IV. Science and Society

## FUNDING FOR RESEARCH AND DEVELOPMENT

Science's major product is an understanding of how things work in the physical, social, and biological worlds. This product by itself yields little income for scientists, and yet it can be quite costly to attain. Scientists, therefore, are continually dependent on various institutions to support their research costs. The institutions approached (e.g., National Institutes of Health, nonprofit foundations such as the Ford Foundation, industry and university research funds) vary in their preferences, and the reasons offered for requesting research support vary according to what a scientist wants to study and what preferences have been expressed by a potential supporting institution. Sometimes the same support request will be presented to two different institutions, with separate arguments tailored to appeal to each institution's specific interests.

Scientists need financial support for themselves and for equipment, assistants, and other technical research needs. Research costs vary from field to field and among projects. High-energy physics research requires very expensive equipment. For example, a particle accelerator may cost several hundred million dollars to build, and a single experiment using that accelerator may also require a bubble chamber costing several million dollars. On the other hand, much mathematical research is done with a pencil and paper and therefore requires little money for special equipment.

Salary, a fundamental cost, is determined by the researchers' base salaries and the percentage of their time to be spent on the project. Salaries vary from field to field (in salary terms it is much better to be a computer scientist than a linguist) and from employer to employer. The self-employed earn the most (an average of $18,000 yearly); those who work for state and local governments, the least ($11,200) (National Science Foundation, 1967). In general, scientists are well-paid professionals whose salaries constitute a major cost of research.

Anyone who hires a scientist is hiring expensive talent and must usually pay, in addition, for expensive equipment and assistants. For this considerable investment one might expect an employer to be guaranteed beneficial results. However, the success of a scientific project cannot be guaranteed. A study of 45 research and development (R&D) projects in one firm showed that only 44 percent of these projects were successful (Mansfield, 1968:50). There is good evidence that this laboratory's experience is not unusual.

A scientist must eat, however, so he must be paid by someone either as a scientist or in some other capacity. (Einstein worked as a patent clerk.) Many scientists are employed as teachers who, as part of their contracts, have some time set aside for research. For example, a scientist may be granted all or part of a term for research, and some scientists, as was noted in the work style discussion, are hired purely as researchers.

A review of the practical results of research entitled *Technology in Retrospect and Critical Events in Science (TRACES;* Leollbach, 1968) highlights the difficulty that scientists experience in justifying basic scientific research as opposed to development work. The report distinguishes three kinds of research: (1) Nonmission (basic) research is "research motivated by the search for knowledge and scientific understanding without special regard for its application"; (2) Mission-oriented (applied) research is performed to develop a particular application; and (3) Development and application is the production of prototypes and engineering design (Leollbach: ix).

*TRACES* reports on five major technological innovations of some use to society. They are: magnetic ferrites for computer memories, video tape recorders, the oral contraceptive pill, the electron microscope, and matrix isolation for chemical processing. It reports the following paraphrased from Leollbach (pp. iv–v).

In all cases studied, nonmission research provided the origins from which science and technology could advance. Of the key events in these innovations, 70 percent were nonmission events, 20 percent mission-oriented, and 10 percent development and application. Table 9 shows where the key events in these innovations took place with respect to universities and colleges, research institutions and government laboratories, and industry. Clearly, universities generally do nonmission research while industry does most of the mission and development research.

Most nonmission research used in a technological innovation takes

## Table 9. The Distribution by Performers of Key Events

| | University and College | Research Institute and Government Labs | Industry |
|---|---|---|---|
| Nonmission research | 76% | 14% | 10% |
| Mission-oriented research | 31 | 15 | 54 |
| Development and application | 7 | 10 | 83 |

*Source:* Leollbach, 1968, p. iv.

place between twenty and thirty years *before* the innovation, and 90 percent has been completed ten years before the innovation is generally available. The mission-oriented and development work occur during the decade just prior to the innovation's availability. The development of an innovation from conception to demonstration of its workability took an average of nine years in these five cases. In summary, then, innovation for our children—from birth control to pollution control—depends on *today's* basic science.

Figure 1 illustrates this truth by outlining the development of the oral contraceptive pill. This figure shows four strands that contributed to the final development: the need for contraception, study of reproductive physiology, hormone research, and steroid chemistry. The general pattern of discovery is shaped by the way the study was done. When we trace a discovery backwards to its inception as an idea, we generally find more than one item leading to that particular discovery. If we take each subdiscovery in turn, we generally find that each has several "parents," branching like a family tree. By emphasizing the major events leading to this discovery, *TRACES* poses two problems:

1. How much of science, at any one time, will result in useful products at some time in the future?
2. To how many other innovations (beyond a specific goal) will one discovery lead?

We know that many scientific discoveries do not result in any use. We also know that some discoveries have several end-results. Figure 1 should indicate the kinds of problems which exist when we try to assign a value to any particular piece of basic research.

Despite the failure rate and the fact that many "successful" projects are not profitable in terms either of company profits or agency mission, science receives substantial support. Table 10 shows transfers of funds from one part of the American economy to another for research and development work in the year 1970. It shows that the federal government, which supports most research, spent a total of $14.7 billion for research.

## Table 10. 1970 Transfers of Funds

**R and D Performances By Sector**

| Sources of R and D Funds | Federal Government | Industry | College and University | Other Non-profits | Total |
|---|---|---|---|---|---|
| Federal Government | 3,876 | 7,784* | 2,395* | 650* | 14,705 |
| Industry | — | 10,074 | 62 | 90 | 10,226 |
| College and University | — | — | 970 | — | 970 |
| other Non-profits | — | — | 166 | 220 | 386 |
| Total | 3,876 | 17,858 | 3,593 | 960 | 26,287 |

*Source:* National Science Foundation, 1972, p. 25.

\* Includes federally funded research and development centers.

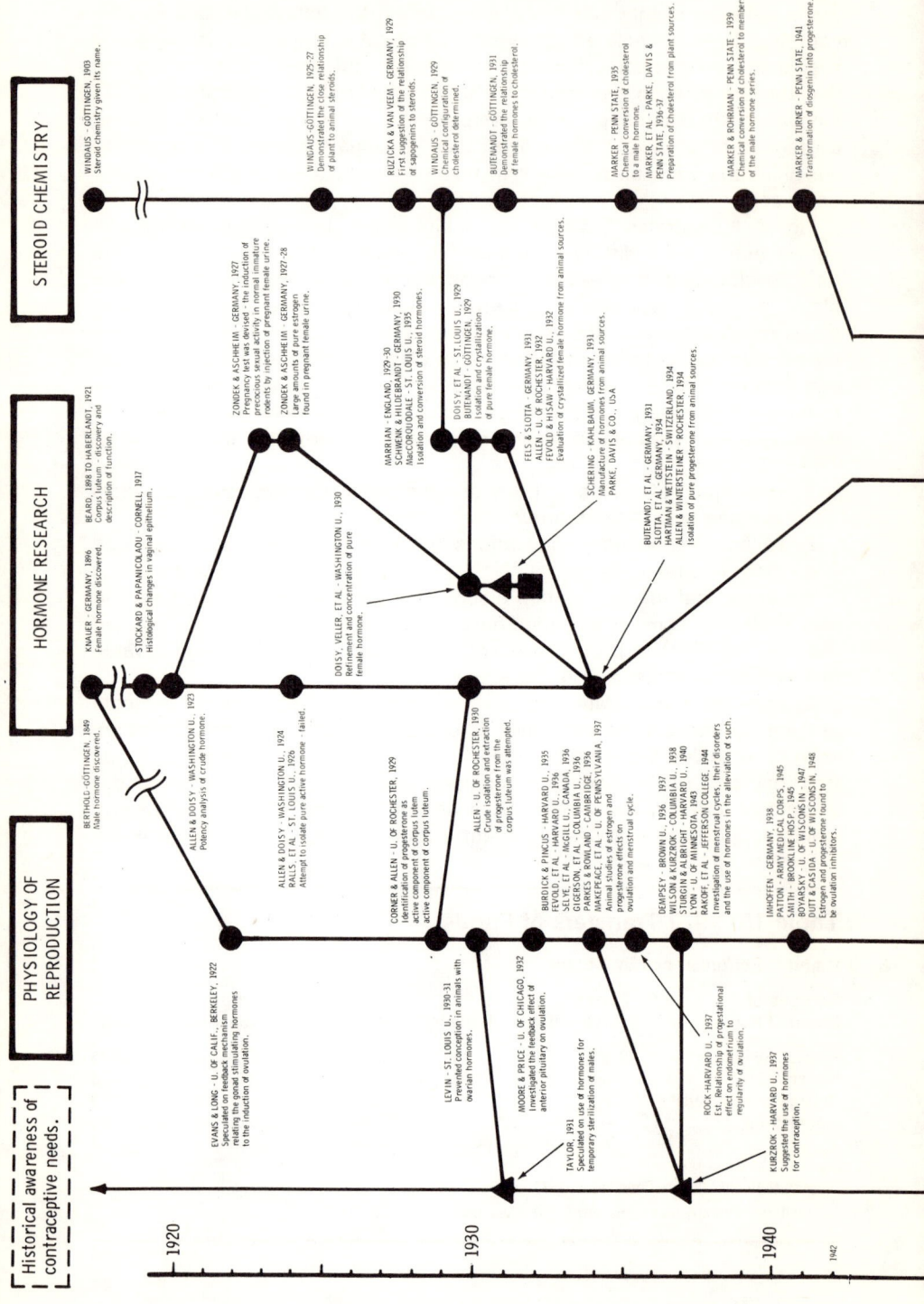

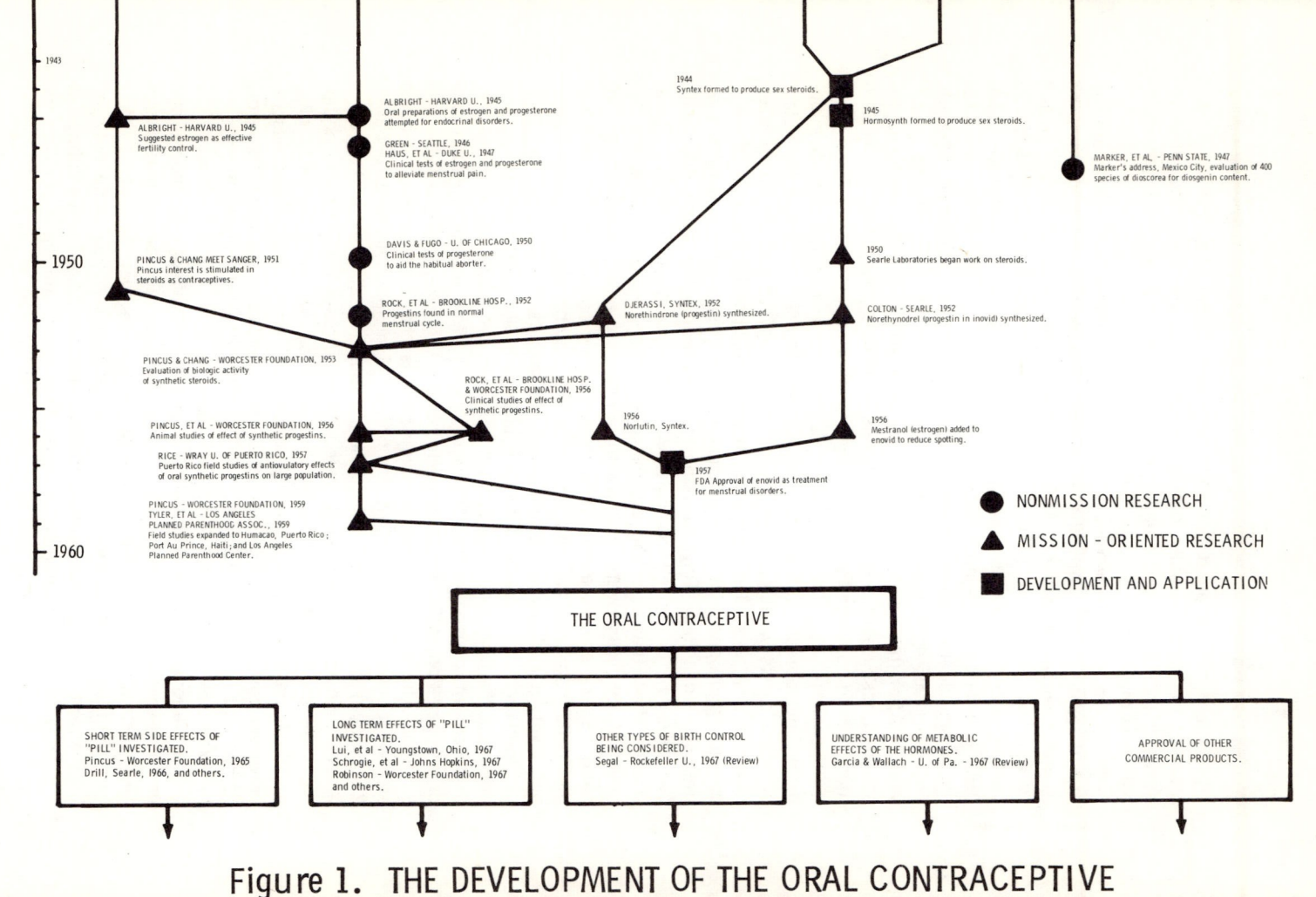

Figure 1. THE DEVELOPMENT OF THE ORAL CONTRACEPTIVE

Of this total, $3.9 billion was spent in federal installations, $7.8 billion in industry, $2.4 billion in colleges, etc. Industry provides the next largest amount but spends almost all of it internally. Colleges spend their total research budgets internally, but they take in a larger percentage than industry (although a much smaller absolute amount of their total research bill) from other sources, largely the federal government.

Within the federal government the largest spenders (as of 1970) are the armed forces, particularly the Air Force. The next two largest are the Atomic Energy Commission (AEC) and the National Aeronautic and Space Administration (NASA). In 1970, these three agencies accounted for 90 percent of all government expenditures on science. The cutbacks in NASA, AEC, and defense research expenditures during the late 1960s have produced significant changes in the amount of funds available to do research. The year 1965–1966 marked a peak in governmental expenditures for research. After this point, a desire for federal economy either reduced research funds or held them at prior levels, even though, as noted earlier, the funds had to be spread among a larger number of scientists.

These large quantities of money support both research and development. Development is the technological end of science, and a persistent question exists whether science and technology should be categorized together. Several studies have been made on this point because much support for (and opposition to) science has been based on end results, e.g., from some pacifists: "Physics made the atom bomb; therefore, don't support physics" (or the reverse statement from some military hawks). These studies show that scientific discovery and the uses of science are not identical. Even discovery alone is not an isolated affair. Scientists do not build a space ship or a new vaccine by starting anew each time. Rather, they build on previous results, some of which may be very old, some newer, and a very few, if any, quite new basic discoveries.

The dilemma of science is that it is useless in the short run but very important in the long. Yet most support decisions are made with respect only to short-term consequences; thus support for basic research, which may or may not "pay off," is very hard to obtain. One example illustrates this situation. The National Science Foundation is chartered to support basic science, but the NSF budget totalled only $525 million in fiscal 1970, barely 2.5 percent of the federal government's total research expenditures. In industry, almost no money is allocated for basic research. Funds for equipment are granted largely on the terms of those who have funds to give, and their basic purpose is clearly to promote technology. Scientists receive support for teaching and other jobs, but many fields are severely hampered by a lack of equipment with which to do basic research (physics, for example—see National Academy, 1965).

The National Science Foundation examined the effect on colleges and universities of changes in federal funding policies made between 1968 and 1970 (*Science,* 1970: 609–612). The findings show that:
1. Although *total* support for colleges and universities is up between seven and eight percent for the two-year period, effective support is down five to ten percent because of inflation and rising enrollments.

2. States are increasing their support of public colleges, but private ones are in trouble.
3. The number of appointments to scientific positions continues to increase but at a slower rate than the number of graduates eligible for those jobs.
4. The federal government is dropping its support of graduate students at a rapid rate, but total enrollments are dropping in only a few areas (e.g., mathematics).

## SCIENCE AND THE POLITICAL ORDER

Any group that requests money from the federal government must defend its share against others who also want money. A science affairs community has been formed by scientists with experience in the federal jungle (e.g., James Shannon, former head of the National Institute of Health, and Philip Handler, National Academy of Science president in 1971). Its strength has been fairly considerable because science's concerns have often coincided with the political needs of other powerful groups such as the military and the medical profession (see, for example, Dupre and Lakoff, 1962).

The federal government's science advisory system has several loosely connected parts. The president has a science advisor who also heads the Office of Science and Technology which reports to the president's office. There are science advisory committees to practically every organization in the government from the White House (the President's Science Advisory Council—[PSAC]) to the Department of Defense and the Bureau of Fisheries; there are scientists working within the government for the many scientific laboratories and organizations (e.g., National Institutes of Health or Weather Bureau). Finally, there is the National Academy of Science with its three parts: the academy itself, the National Research Council, and the National Academy of Engineering. The National Academy of Science was chartered by Congress in 1863 to advise the federal government on scientific matters.

*Science,* the official journal of the American Association for the Advancement of Science, is a weekly magazine which reports the activities of the science advisory structure outlined just above and their relation to the course of national policy and the other parts of government.[12] For example, on October 23, 1970, *Science* outlined the process by which a new science advisor to the president had been selected, highlighting the changes in the selection process and speculating on why those changes had occurred. *Science* indicated how much interaction exists between the various sources of potential science funding and how much of that funding depended on the political process. Another news item in the same issue, entitled "Cyclamates: House Report Changes Administrative Alchemy at

12. *Science* contains several parts. It is a general interest journal containing three or four major articles a week on a wide range of topics. It is also one of the more respected biomedical journals. Its "News and Comments" section acts as a weekly newsmagazine for science. *Science* is a necessity for the person who wants to know what is happening in the scientific community at any given time. It is available in most college libraries.

HEW," discussed a House of Representatives report on the banning of cyclamates in food products. It detailed the activities of the secretary of Health, Education and Welfare (HEW) and the Food and Drug Administration during the fall of 1969 when this first major ban of a food product occurred. A third item reported the difficulties of Professor Jack Kurzwell in retaining his job at San Jose State College. The federal government's impact on colleges was the theme. Dr. Kurzwell is married to a Communist, and this fact seemed to be the source of his problem. The final news discussed was the 1970 Nobel Prizes in Medicine and Physiology.

The political process that leads to funding has been the subject of many studies in the recent past. Committees (generally sponsored by the National Academy of Science) have investigated various parts of science, such as physics (1965) and the social sciences (1969), and in the process have produced voluminous reports generally stating that the particular science needed more money. These reports were the subject of a biting conclusion by D. Greenberg (1970), *Science* news editor at the time the following was written:

> *Science policy ruminating has developed into an academic cottage industry, one that even provides its grave practitioners with opportunities to appear as witnesses on Capitol Hill, which, no doubt for many of them, is a highly esteemed relief from the obscurity of the campus. However, the relation between their prodigious output and the actual makings of science policy or practice appears to be in the vicinity of nonexistence. The pastime is harmless enough and may even be justified as a dignified diversion for those who have tired of the real business of research or research administration. But if the scientific community is as financially malnourished as it claims to be, and if the consequence of this state is as disastrous for the nation as science's leaders contend it is, then perhaps it is time to try something besides multi-hundred-page reports that even many of the faithful have tired of reading. For one fact is apparent: political Washington does not share the scientific community's view of the value or plight of science. This being so after long efforts at courteous persuasion, then other means might profitably be considered. Prominent among them, of course, is the use of organizations, money, and votes in behalf of those legislators who are disposed to act on the assumption that the national research enterprise merits additional support. After all, such political action is an old American custom.*

Greenberg's injunction to direct involvement in politics itself seems to have been taken seriously only by those *not* presently holding top roles in science.

Scientists are, however, involved in the political process as advisors on such questions as environmental pollution, population control, and weapon systems design and adequacy (e.g., Art, 1968). Direct political action in the sense of setting up committees to support friendly congressmen reduces scientists to the role of a small, quaint pressure group with very little influence. By contrast the impact of a top scientist advising on policies with a large scientific component such as the long-term effects of low-grade radiation (a major problem in the development of nuclear power plants) is tremendous.

When we consider such issues and the place of scientists in them, we must remember that there is rarely a clear-cut "scientific" answer to the policy question being asked. The scientist must make assumptions (e.g., of the improvements in shielding from low-level radiation that are likely to be made in the future), estimate unknowns (e.g., the effects of long-term, low-level radiation have not been conclusively investigated), and assess probabilities (e.g., how likely is an accident?) in order to provide a series of answers which decision makers and the public can use. Yet the decision to build power plants is being made now. These assumptions, estimations and assessments are the sources of disagreement among scientists on policy issues. Much of the outcry about decisions such as those on nuclear power plant construction can be traced to the hidden nature of this advice (which leaves groups opposing governmental policies at a disadvantage) (Von Hippel and Primack, 1970) and the inability of either decision makers or their opponents to understand the bases of the advice that is being given.

The science policy cottage industry will probably have some very interesting occurrences to observe in the near future as environment, privacy, productivity, and other problems with high scientific and technological components are taken up by policy makers of all varieties. Scientists will not speak unanimously on these issues, and matters of scientific judgment will become matters of public debate.

It seems clear that, where appropriate, scientists must use their knowledge to affect political decisions, partly because their expertise is needed, partly because a proposal is helped if that man in his white coat is extolling the virtues of this particular new way to solve a political problem. This need indicates the increased importance of scientific issues as well as the increased number of scientists. And the effectiveness of science's symbols has been increased by our environment in which more and more people have been exposed to science at an elementary level.

The scientific community will continue to probe issues that have little or nothing to do with any matter of public policy but much that is important to the intellectual structure of science. In the activities of the scientific community we will continue to find ideas that excite men to work endless hours under difficult conditions in order to understand them. The people within this community will admit that this exhausting work is tremendously exhilarating.

How many scientists will receive basic salaries and research support only as scientists is a question that must be answered by those who have support to give. Science will continue to be done whether or not research support is given. When funds are denied to science, however, it is always the developmental scientists who suffer the most because their claims for their projects are subject to direct investigation. For example, in the period between 1967 and 1972, when federal funding for science was slowed, the support for basic research increased by 23 percent, for applied research by 27 percent, and for development by only 15 percent. This slowdown, combined with inflation which has made the research dollar worth less, has caused many of the difficulties mentioned above.

# Conclusion

We have examined four aspects of science: (1) the scientific occupation, community, and work style; (2) the scientific career; (3) the social organization of science; and (4) science and the political order. We have seen that the occupation of "scientist" is both well populated and growing. The rate of growth over the past twenty-five years will clearly not continue, and the future growth rate is still an open question. We distinguished between occupation per se and the active scientific community. The active group is central to communication, creative research activity, and training of other scientists.

The scientific career results from a series of successes in academic competition. Potential scientists are recognized early, and their numbers are reduced by each higher level of education. The final sort is that which separates active scientists from the total in scientific occupations. This career selection system has been built within the standard, American competitive education system, and its continued existence may be threatened as educational forms change. Questions of existence aside, however, the career system is also in trouble (although the full seriousness is not evident yet) because the supply of scientists is rapidly outrunning the number of jobs available. Unemployment, which has affected all levels of technical employment, has made scientific occupations less attractive. The effect of this change on the number of new scientists could be severe since those who decide to leave the potential scientists' pool rarely return. Do we indeed have enough scientists that a substantial drop is desirable? Is it desirable that our system of professionalization results in only Ph.D.'s doing research? Do we perhaps need more scientists but many with less than Ph.D. training? All are open questions which need answers before we can evaluate the change in our system for producing scientists.

The social organization of science is built around oral and written communication which is highly redundant. Many attempts have been made to "rationalize" the communication structure by reducing the redundancy, but resistance to these attempts by scientists indicates how important the redundancy is to the social structure. Changes in communication structures can produce changes in the structure of disciplines and specialties and eventual reorganization of the scientific community.

Finally, we examined science and society. Science is supported by both governmental and nongovernmental sources even though it has no record of immediate payoff for support. (Indeed, basic science may be decades away from concrete usage, if any concrete usage at all is found.) This support is not automatic, and science has had to struggle and lobby to gain support for itself, particularly from the federal government. Scientists' most effective political activity is advising the federal government on matters of policy which have technical elements.

On balance, science itself will continue to be done even though many elements traditional to the institution of science are in a state of flux.

# References

**American Psychological Association**
1963 "Project on Scientific Information Exchange in Psychology." Reports 1, December, Washington, D.C.

**Art, Robert J.**
1968 The TFX Decision, McNamara and the Military. Boston: Little, Brown and Co.

**Barber, Bernard**
1966 Science and the Social Order. New York: Free Press.

**Barber, Bernard and Walter Hirsch, eds.**
1962 The Sociology of Science. Glencoe: Free Press.

**Ben-David, Joseph**
1971 The Scientist's Role in Society: A Comparative Study. Englewood Cliffs, N.J.: Prentice-Hall.

**Ben-David, Joseph and R. Collins**
1966 "Social factors in the origins of a new science: the case of psychology." American Sociological Review 31: 451–465.

**Bennis, Warren G.**
1956 "Values and organization in a university social research group." American Sociological Review 21: 555–563.

**Cairns, John, Gunther S. Stent, and James D. Watson**
1966 Phage and the Origins of Molecular Biology. Cold Spring Harbor, N.Y.: Cold Spring Harbor Laboratory of Quantitative Biology.

**Cole, Stephen and Jonathan R. Cole**
1967 "Scientific output and recognition: a study in the operation of the reward system in science." American Sociological Review 32: 377–390.

**Cooley, William W.**
1958 "The application of a development rationale and methods of multivariate analysis to the study of potential scientists." Unpublished doctoral dissertation. Harvard University.

**Cordtz, Dan**
1971 "Bringing the laboratory down to earth." Fortune 83(1): 107.

**Dupre, J. Stefan and Sanford A. Lakoff**
1962 Science and the Nation: Policy and Politics. Englewood Cliffs, N.J.: Prentice-Hall.

**Eisenstadt, S. N.**
1964 From Generation to Generation: Age Groups and Social Structure. New York: Free Press.

**Evans-Pritchard, Edward Evan**
1940 The Nuer. Oxford: Oxford University Press.

**Fisher, C. S.**
1966 "The death of mathematical theory: a study in the sociology of knowledge." Archives for History of Exact Sciences 3(2): 132-159.

1967 "The last invariant theorists." European Journal of Sociology 8(2): 1, 216–244.

**Fleming, Donald**
1969 "Emigré physicists and the biological revolution." In Donald Fleming and Bernard Bailyn, The Intellectual Migration—Europe and America 1930–1960. Cambridge, Mass.: Harvard University Press. Pages 152–189.

**Glaser, Barney G.**
1964 Organizational Scientists: Their Professional Careers. Indianapolis: Bobbs-Merrill.

**Greenberg, Daniel**
1970 "Academic research: OST aide sees no shift in financial situation." Science 170 (November 27): 154.
**Hagstrom, Warren O.**
1965 The Scientific Community. New York: Basic Books.
1968 "Departmental prestige and scientific productivity." Paper read at the meeting of the American Sociological Association, August, Boston.
**Hall, A. R.**
1963 "Merton revisited, or science and society in the seventeenth century." History of Science 2: 1–15.
**Hare, Paul A.**
1962 Handbook of Small Group Research. New York: Free Press.
**Hirsch, Walter**
1968 Scientists in American Society. New York: Random House.
**Kaplan, Norman, ed.**
1965 Science and Society. Chicago: Rand McNally.
**Koestler, Arthur**
1964 The Act of Creation. London: Hutchinson.
**Kuhn, Thomas**
1970 The Structure of Scientific Revolution, 2d ed. Chicago: University of Chicago Press.
**Leollbach, Herman, ed.**
1968 Technology in Retrospect and Critical Events in Science (TRACES). Chicago: IIT Research Institute.
**Mansfield, E.**
1968 Industrial Research and Technological Innovation: an Econometric Analysis. New York: W. W. Norton.
**Martino, Joseph P.**
1969 "Science and society in equilibrium." Science 165 (August 22): 769–772.
**Menzel, Herbert**
1962 "Planned and unplanned scientific communication." In the Sociology of Science, ed. Bernard Barber and Walter Hirsch. Glencoe: Free Press.
**Merton, Robert**
1938 "Science, technology, and society in seventeenth century England." Osiris 4 (Part II): 360–632. Reprinted, 1970, New York: Harper Torchbooks.
1959 "Priorities in scientific discovery." American Sociological Review 22 (December): 635–659.
1961 "Singles and multiples in scientific discovery." Proceedings of the American Philosophical Society 105 (October): 470–486.
**Mullins, Nicholas C.**
1966 "Social networks among biological scientists." Unpublished doctoral dissertation. Harvard University.
1972 "The development of a scientific specialty: the phage group and the origin of molecular biology." Minerva X (January): 51–82.
**National Academy of Sciences—National Research Council Committee on Science and Public Policy**
1965 Physics: Survey and Outlook. Washington, D.C.: National Academy of Sciences.
1969 The Behavioral and Social Sciences: Outlook and Needs. Englewood Cliffs, N.J.: Prentice-Hall.
**National Science Foundation**
1967 American Science Manpower, 1966: a Report of the National Register of Scientific and Technical Personnel. Washington, D.C.: U.S. Government Printing Office. NSF 68–7.
1969 Annual Report 1968. Washington, D.C.: U.S. Government Printing Office. NSF 69–1.

1971 American Science Manpower 1970: a Report of the National Register of Scientific and Technical Personnel. Washington, D.C.: U.S. Government Printing Office. NSF–71–45.
1972 National Patterns of R & D Resources. Washington, D.C.: U.S. Government Printing Office. NSF 72–300.

**Nelson, William R., ed.**
1968 The Politics of Science. New York: Oxford University Press.

**Newsweek**
1970a "Doctoral glut," March 16, page 114.
1970b "Supply and demand: teacher glut," June 29, pages 58–59.

**Organisation for Economic Cooperation and Development (OECD)**
1968 Reviews of National Science Policy. United States.

**Pelz, Donald C. and Frank M. Andrews**
1966 Scientists in Organizations: Productive Climates for Research and Development. New York: John Wiley and Sons.

**Peter, Laurence J. and Raymond Hull**
1969 The Peter Principle. New York: Bantam Books.

**Popper, Karl R.**
1959 The Logic of Scientific Discovery. London: Hutchinson.

**Price, Derek deS.**
1963 Little Science, Big Science. New York: Columbia University Press.

**Price, Don K.**
1965 The Scientific Estate. Cambridge: Belknap Press of Harvard University Press.

**Reif, F.**
1961 "The competitive world of the pure scientist." Science 134 (December 15): 1957–1962.

**Roe, Anne**
1953 The Making of a Scientist. New York: Dodd, Mead and Co.

**Science**
1970 "News and comments." 170 (November 6): 609–612.

**Snow, C. P.**
1959 The Two Cultures and the Scientific Revolution. Cambridge: Cambridge University Press.
1961 Science and Government. Cambridge, Mass.: Harvard University Press.

**Statistical Abstract of the United States**
1969 Washington, D.C.: U.S. Government Printing Office.

**Storer, Norman W.**
1966 The Social System of Science. New York: Holt, Rinehart and Winston.

**U. S., President's Science Advisory Committee**
1968 Science, Government and Information. Washington, D.C.: U.S. Government Printing Office.

**Von Hippel, F. and J. Primack**
1970 The Politics of Technology. Palo Alto, Calif.: Stanford Workshop on Political and Social Issues.

**Watson, James D.**
1968 The Double Helix: a Personal Account of the Discovery of the Structure of DNA. New York: Atheneum.

**Wilson, K. M.**
1965 Of Time and the Doctorate. Atlanta: Southern Regional Educational Board Research Monograph 9.

**Ziman, John**
1968 Public Knowledge: The Social Dimension of Science. Cambridge: Cambridge University Press.

**Zuckerman, Harriet A.**
1967 "The sociology of the Nobel Prizes." Scientific American 219(5): 25–33.

# Suggested Readings

The following selections are recommended as clear and interesting presentations of the sociology of science.

**Ben-David, Joseph**
1971   The Scientist's Role in Society. Englewood Cliffs, N.J.: Prentice-Hall.   Ben-David is the master of the historical development of science over the period since the Middle Ages. His view of the development of science as an institution and of its place in universities is an important contribution to the sociology of science.

**Merton, Robert**
1971   Science, Technology and Society in Seventeenth Century England. New York: Harper Torchbooks. This book contains a reprinting of Merton's 1938 essay on science as it developed as a social institution in seventeenth-century England. His views complement Ben-David's later work. Taken together the two provide a comprehensive view of the institution of science.

**Kuhn, Thomas**
1970   The Structure of Scientific Revolutions, 2d ed. Chicago: Univ. of Chicago Press.   This work is the beginning of a whole tradition of thought about science. His insights into the way that scientific ideas change have been very important to contemporary thinking about how science is done.

**Price, Derek deS.**
1963   Little Science, Big Science. New York: Columbia Univ. Press.   Price is a historian of science whose contributions to sociology have been enormous. His book is a collection of essays about science in general; it suggests the "invisible college" concept, asks how smart and how productive the average scientist is, and makes interesting observations about how science works.

**Zuckerman, Harriet**
1971   "Stratification in American science." In Social Stratification: Research and Theory for the 1970s, ed. Edward O. Laumann. Indianapolis: Bobbs-Merrill. This article reviews material on the causes and effects of the fact that some scientists are very well known in science but that most are totally unknown —a situation resulting from the distribution of productivity that Price discusses in Little Science, Big Science. Zuckerman covers many studies done by the Columbia group of sociologists of science.

**Watson, James D.**
1968   The Double Helix. New York: Atheneum Press.   This book is an autobiography which describes an important scientific discovery. Watson is neither modest nor forgiving of others, but he is entertaining and true to one kind of scientific life: research in a competitive field at the very top of its creativity.

# THE BOBBS-MERRILL REPRINT SERIES

The author recommends for supplementary reading the following related materials. Please fill out this form and mail.

*Indicate number of reprints desired*

____ **Ben-David, Joseph and Awraham Zloczower** 1962 "Universities and academic systems in modern societies." European Journal of Sociology, pp. 48–84. **S-547**/66924 60¢

____ **Heilbron, John L. and Thomas S. Kuhn** 1969 "The genesis of the Bohr atom." Historical Studies in the Physical Sciences, pp. 211–290. **HS-28**/68466 $1.00

____ **Kuhn, Thomas S.** 1959 "Energy conservation as an example of simultaneous discovery." In Critical Problems in the History of Science, ed. Marshall Clagett. University of Wisconsin Press. **HS-42**/68480 60¢

____ **Merton, Robert K.** 1957 "Priorities in scientific discovery: a chapter in the sociology of science." American Sociological Review, pp. 635–659. **HS-54**/68492 60¢

____ **Rosenberg, Charles E.** 1967 "Factors in the development of genetics in the United States: some suggestions." Journal of the History of Medicine and Allied Sciences, pp. 27–46. **HS-63**/68500 40¢

____ **Truesdell, C.** 1968 "A program toward rediscovering the rational mechanics of the Age of Reason." In Essays in the History of Mechanics by C. Truesdell, Springer-Verlag, pp. 84–137. **HS-76**/68513 80¢

____ **Zuckerman, Harriet A.** 1967 "Nobel laureates in science: patterns of productivity, collaboration, and authorship." American Sociological Review, pp. 391–403. **S-655**/67032 40¢

*The Bobbs-Merrill Company, Inc.*
*College Division*
*4300 West 62nd Street*
*Indianapolis, Indiana 46268*

Instructors ordering for class use will receive *upon request* a complimentary desk copy of each title ordered in quantities of 10 or more. Refer to author and *complete* letter-number code when ordering reprints.

☐ Payment enclosed  ☐ Bill me (on orders for $5 or more only)

_____ Course number  _____ Expected enrollment

☐ For examination  ☐ Desk Copy

Bill To _____

ADDRESS _____

CITY _____ STATE _____ ZIP _____

Ship To _____

ADDRESS _____

CITY _____ STATE _____ ZIP _____

Please send me _____ copies of the sociology reprints catalog.

Please send me related reprints catalogs in _____

Any reseller is free to charge whatever price he wishes for our books.

*For your convenience please use complete form when placing your order.*